ISBN: 978-1-5272-1225-1

Typeface: Futura and Calibri within body, Trajan and Trebuchet on front and back covers

Further copies obtainable from Paul Nichol
E-mail: pnichol20@gmail.com

WEa
EAST MIDLAND

First published by the Workers Educational Association,
East Midlands Region
All other editions published by Paul Nichol

Table of Contents

Table of Images

- Cover image: with permission Tony Marsh
- Page 7, lower right: Wikipedia, Creative Commons Benjamin Dion, see also Image Number 395215 at http://mushroomobserver.org/
- Page 7: all other images: Paul Nichol
- Page 9: with permission Joe Gilbert
- Page 10, upper left: Wikipedia, Creative Commons Sasata
- Page 10, upper right: with permission Andrew Denley
- Page 10, lower left: Wikipedia, Creative Commons H. Krisp
- Page 10, lower right: Wikipedia, Creative Commons Th. Kuhnigk
- Page 11, all images: with permission Andrew Denley
- Page 12, upper left: Wikipedia, Creative Commons Tony Hisgett
- Page 12, all other images: with permission Andrew Denley
- Page 13, upper left: Wikipedia, Creative Commons David Perez
- Page 13, upper middle: Wikipedia, © 2008 Dan Molter, see also Image Number 27365 at http://mushroomobserver.org/
- Page 13, all other images: with permission Andrew Denley
- Page 14, upper: Wikipedia, Creative Commons Dejan Stojanovic, https://www.flickr.com/photos/fungi-nis/5773219663 (no changes made)
- Page 14, lower: Wikipedia, Creative Commons Jason Hollinger, see also https://www.flickr.com/photos/7147684@N03/4037847776/
- Page 15: with permission John Harris, www.mushroomdiary.co.uk
- Page 16, both images: with permission Andrew Denley
- Page 17, both upper images: with permission Andrew Denley
- Page 17, bottom image: Wikipedia, Creative Commons Zdenek1945, Czech Republic

- Page 18: Wikipedia, Creative Commons H. Krisp
- Page 19: Wikipedia, Creative Commons James Lindsey, see also www.commanster.eu/commanster
- Page 22: with permission Andrew Denley
- Page 23, left: Wikipedia, Creative Commons Alan Rockefeller
- Page 23, upper right: Wikipedia, Creative Commons Tatiana Bulyonkova, see also www.flickr.com/photos/ressaure/15269601422
- Page 23, lower right: with permission Andrew Denley
- Page 26: with permission Tony Marsh
- Page 35: with permission Andrew Denley

Acknowledgments

Thanks to all the users of these keys who continue to make useful suggestions for improvements. This support has been valuable, allowing me to redesign the pathways in some of the keys. Thanks to my daughter Thea for providing sketches, which appeared in earlier editions, and now on page 5, and to Andrew Denley for allowing his superb photographs of Cumbrian fungi to be reproduced. Thanks also to Joe Gilbert for providing photographs of the student groups enjoying their time amongst the fungi, and to John Harris for the Gill drawings. Finally, thank you to my son-in-law Brent for reworking the design and layout of the revised third edition, the fourth edition and this revised fourth edition.

Introduction

Some of the larger fungi are well known, and by having certain characteristic features are readily identified. To a casual observer however the majority look very much alike and it is often a difficult task to be certain about an identification. This key will hopefully point the interested mushroom collector in the right direction by identifying a fungus to its GENUS, the vital first step to an identification. To be able to identify a specimen to its genus is a worthy accomplishment for any naturalist; to identify further to the species requires more detailed characteristics to be examined which are often microscopic e.g. spore size. It is possible however to identify many species from visible characteristics. Having determined the genus, it then becomes possible to focus attention on only those species that occur in the genus and with the help of some of the excellent texts and Field Guides now available the actual species can be often determined. There remain, however, some genera in which the gross morphology of many of the species is very similar and microscopic features have to be examined.

This key was written in response to a need felt by students on courses organized by the Workers Educational Association for an initial aid to help them develop their skills in identification of fungi collected on forays. Not all the genera to be found in Britain are in the keys. Those included represent genera that students regularly encountered in the field and are taken from their lists.

How to Use The Keys

In order to find the correct key for the specimen under investigation the overall structure must first be examined and nine possibilities are provided (page 10). The shape, structure and texture of the specimen will match one of the descriptions and the relevant key is indicated. Each key consists of a flow diagram in which you should follow the arrows. A series of statements are given and you are asked to answer yes or no following an examination of the specimen. Following the relevant arrow will lead to a further statement and so on until the name of the fungus is reached. Success in using any key increases with experience but depends on careful examination of the features of the fungus as it is quite easy to follow the wrong route and therefore it is essential that the key is read from the beginning.

Key 1 contains the fungi with gills, the typical 'mushroom' of which there are many genera. The first step in dividing up this large group is to know the colour of the spores. Individual spores cannot be seen without a microscope but a mass of many millions of spores can and this mass is usually coloured. In order to obtain a spore mass it is necessary to collect the spores as they are shed from the gills by obtaining a spore print. Two techniques can be used but both depend on the mushroom being mature; young and old fruiting bodies do not shed spores.

TECHNIQUE ONE

Cut the cap off the stalk as near to the top as possible without damaging the gills and place it on either a piece of glass or card which is half-black and half-white. If using card ensure that the cap is on both the black and white sides as dark spores will only show on the white card while pale or white spores will be visible on the black. Place a jam jar or similar container over the cap and leave for a few hours, after which time spores will be shed forming a print, the colour can then be determined. If using glass, scrape the spores together into a small pile.

TECHNIQUE TWO

Taking a recently collected entire fruit body push the stalk through a hole in the black/white card so that the underside of the cap rests on the card. Stand the card on top of a glass filled with water and leave until the spores are shed. With the stalk in water the fruit body will retain sufficient moisture which is necessary to liberate the spores.

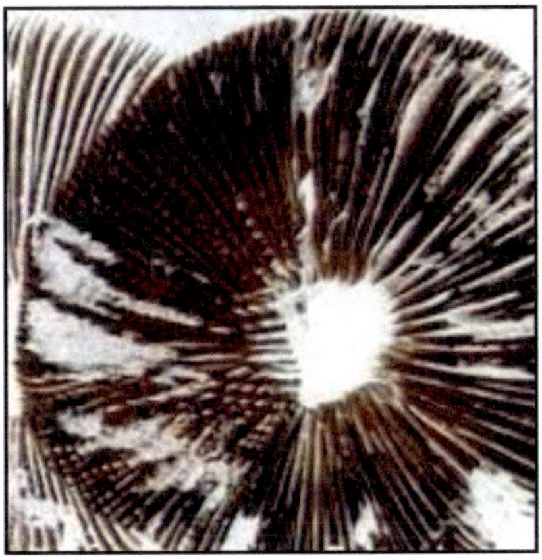

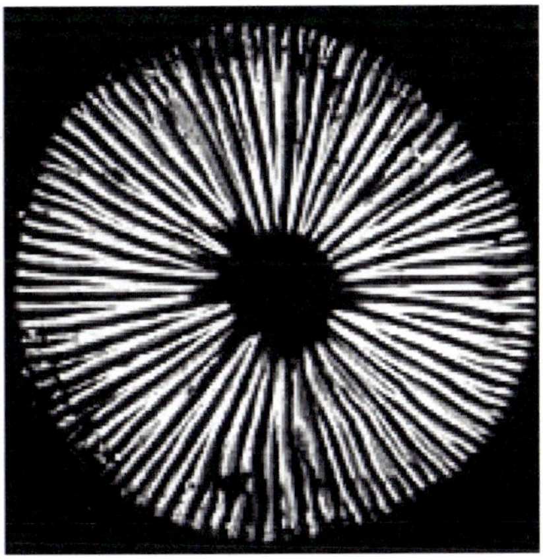

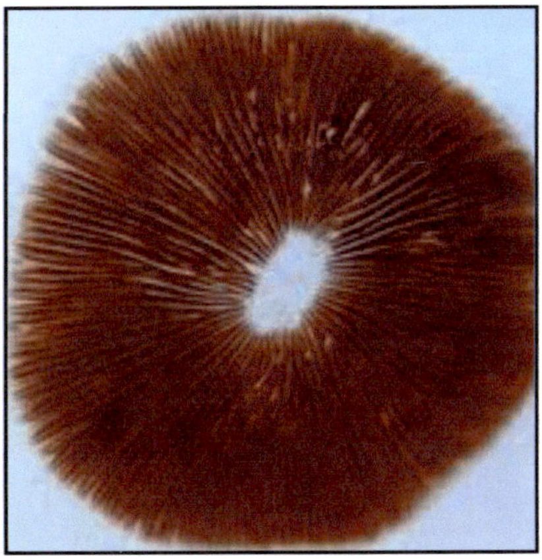

Glossary of Terms Used

Amyloid	A drop of iodine on collected spores and sometimes the gills give a blue / black colour.
Dichotomous	Dividing twice
Fibrillose	Finely fibrous; usually refers to cap or stalk.
Hygrophanous	Cap changing colour depending on whether wet or dry; often exhibiting two different colours.
Intermediate gill	Short gills which do not run from the stalk to the edge of the cap.
Latex	A white but sometimes orange/red liquid emerging when the gills are cut.
Resupinate	Growing flat like a crust; sometimes curled at the edges.
Rhizoids	Thick white cords at the bottom of the stalk.
Ring	Torn remnants of tissue encircling the stalk.
Volva	Tissue that surrounds the base of the stalk like a bag or sheath.

The use of DNA sequencing is resulting in many name changes throughout all branches of natural history. In an attempt to keep up, where we have seen a change in the name of a genus the old name occurs below in brackets ( ).

Index to the keys

KEY 1
Fruit body, a cap with distinct gills or gill-like ridges, with or without a stalk.

KEY 2
Fruit body, a stalk and cap but without gills.

KEY 3
Fruit body, a stalk and head region, not a cap.

KEY 4
Fruit body forming a bracket.

Index to the keys (continued)

KEY 5
Fruit body hemispherical, spherical, or pestle-shaped.

KEY 6
Fruit body forming upright branched or unbranched club-shaped structures.

KEY 7

Fruit body cup-shaped or trumpet-shaped with or without a stalk.

KEY 8

Fruit body forming flat irregular crusts on wood.

KEY 9

Fruit body with a jelly-like or rubbery texture.

KEY 1

Fruit body with some sort of stalk

yes → (see below)

no → *Crepidotus*

Gills shallow or reduced to vein-like ridges, always repeatedly forked

no → Fruit body with distinct gills (see below)

yes ↓

Gills cream/orange regularly dichotomously forked

no ↓

yes → *Hygrophoropsis*

Gills irregularly branched

Stalk hollow → *Craterellus*

Stalk solid → *Cantharellus*

Fruit body with distinct gills

Spores white or cream	KEY 1A
Spores pale to deep salmon pink	KEY 1B
Spores very dark brown, purple or black	KEY 1C
Spores yellow/ brown, rust, or mid brown	KEY 1D

Gill Attachments

Adnate: Gills broadly attached to stalk

Adnexed: Gills narrowly attached to stalk

Decurrent: Gills running down the stalk

Slightly decurrent: Gills running only slightly down the stalk

Free: Gills do not meet the stalk (unattached)

Sinuate: Gills smoothly notched before running slightly down the stalk

KEY 1A Cap with gills: Spores white or Cream

stalk central

no ← fruit body pink or orange - rubbery

- **yes**: *Rhodotus*
- **no** ↓ edge of gills serrated
  - **yes** ↓ smell of aniseed
    - **yes**: *Lentinellus*
    - **no**: *Lentinus*
  - **no** ↓ fruit body large and fleshy
    - **yes**: *Pleurotus*
    - **no**: *Panellus or Sarcomyxa*

yes → with numerous intermediate gills

- **yes** ↓ latex produced
- **no**: *Russula*

- **yes**
  - small delicate types: *Mycena*
  - larger types: *Lactarius*
- **no** ↓ continue to top of 17

continuation of KEY 1A

- with volva
  - yes → *Amanita*
  - no → growing on remains of other fungi
    - yes → *Collybia or Asterophora*
    - no → growing on cones
      - yes
        - in spring → *Strobilurus*
        - in autumn → *Baeospora*
      - no → with ring on stalk
        - no → continue to top of page 18
        - yes → gills slightly decurrent
          - yes → *Armillaria*
          - no → gills free often with a swollen stem base
            - no → cap sticky
              - yes → *Mucidula (Oudemansiella)*
              - no → cap surface powdery → *Cystoderma*
            - yes → fruit body large and robust
              - no → *Lepiota*
              - yes → flesh reddening when cut
                - no → *Macrolepiota or Amanita*
                - yes → *Chlorophyllum*

continuation of KEY 1A

gills decurrent (maybe only slightly)

- **yes** → Fruit bodies growing in tufts or gills bruising blue
  - **yes** → *Lyophyllum*
  - **no** → In grass, fruit body brightly coloured or white
    - **yes** → *Hygrocybe & other waxcap genera*
    - **no** → fruit body small with widely spaced gills
      - **yes** → *Lichenomphalia or Arrhenia*
      - **no** → Cap greasy to slimy
        - **yes** → *Hygrophorus*
        - **no** → stalk base very swollen, club-shaped
          - **yes** → *Ampulloclitocybe*
          - **no** → *Clitocybe or Infundibulicybe*
- **no** → gills free
  - **yes** → stalk brown and velvety → *Flammulina*
  - **yes** → stalk not velvety → *Gymnopus or Rhodocollybia*
  - **no** → gills sinuate
    - **no** → continue to top of page 19
    - **yes** → fruit body large and fleshy
      - **no** → *Mycena*
      - **yes** → gills and flesh yellow
        - **yes** → *Tricholomopsis*
        - **no** → cap flat with a central umbo, spores amyloid
          - **yes** → *Melanoleuca*
          - **no** → thick white rhizoids at base of stalk
            - **yes** → *Megacollybia*
            - **no** → *Tricholoma*

continuation of KEY 1A

cap greasy or sticky

yes → stalk brown and velvety

- yes → *Flammulina*
- no → stalk long, cap fluted
  - yes → *Hymenopellis (Xerula)*
  - no → *Hygrocybe and other waxcap genera*

no → small delicate types with bell-shaped caps

- yes → gills few and widely spaced → *Marasmius or Marasmiellus*
- yes → gills more crowded → *Mycena*
- no → growing in tufts or gills bruising blue
  - yes → *Lyophyllum*
  - no → flesh and gills hard and brittle
    - yes → *Russula*
    - no → gills widely spaced powdered white, cap lilac or pink
      - yes → *Laccaria*
      - no → thick white rhizoids at base of stalk
        - no → *Gymnopus or Rhodocollybia*
        - yes → *Megacollybia*

KEY 1B Cap with gills: Spores pale pink to deep salmon

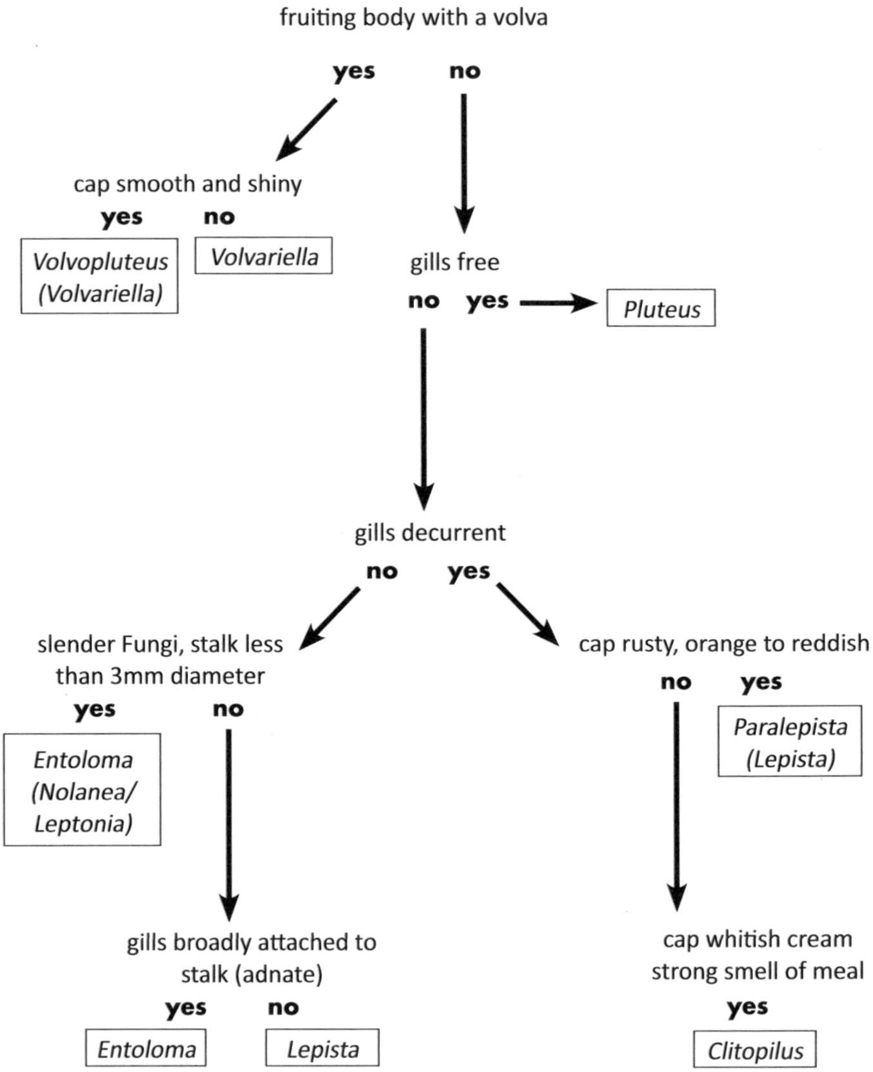

KEY 1C Cap with gills: Spores very dark brown, purple or black

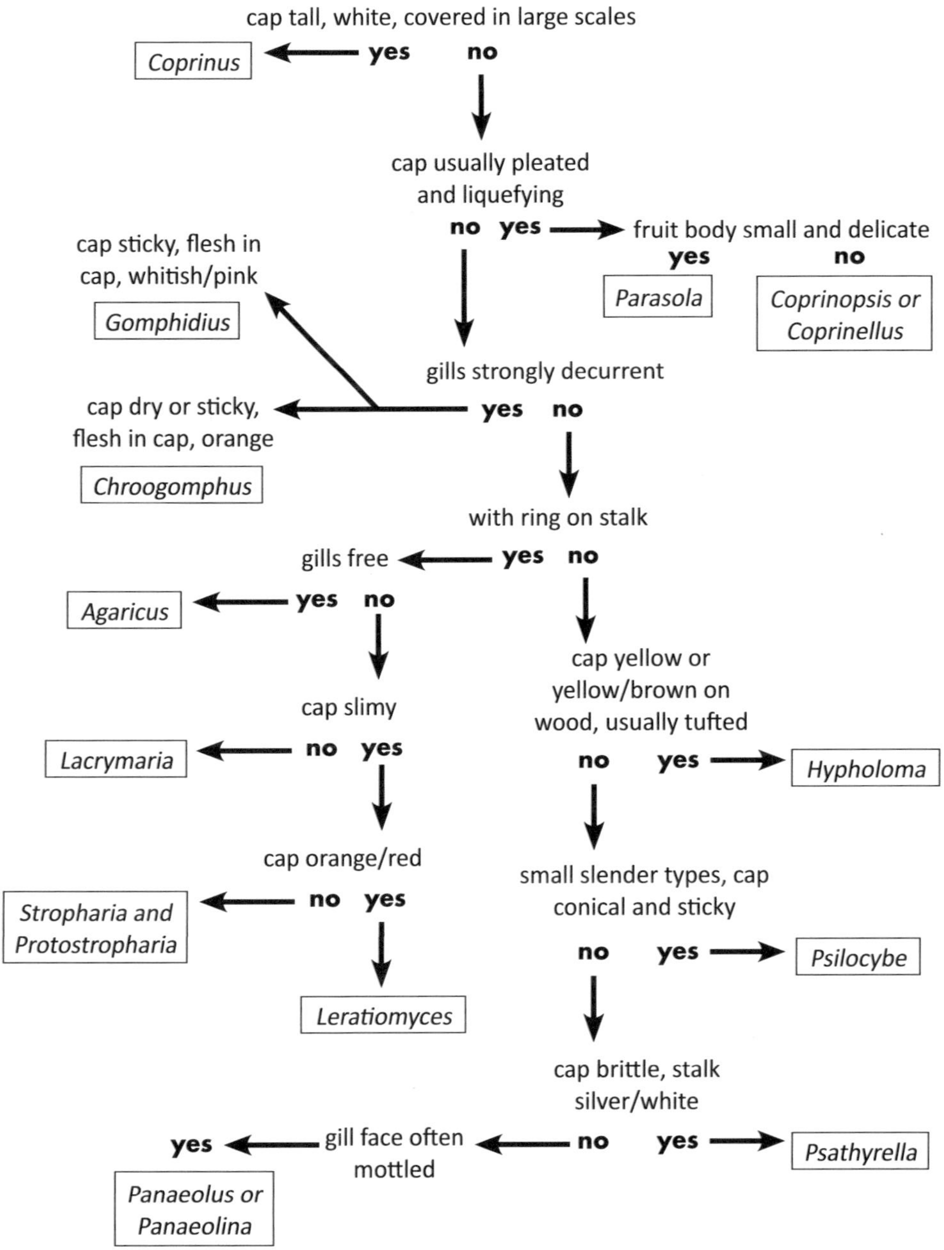

KEY 1D Cap with gills: Spores yellow/brown

fruit body bracket-shaped, if a stalk very stout and velvety
- **yes** → *Tapinella*
- **no** ↓

ring on stalk
- **no** ↓ gills very decurrent
  - **yes** → *Paxillus*
  - **no** ↓ small brownish fruit body, cap flat 1-3cm
    - **yes** → gills slightly decurrent → *Tubaria*
    - **yes** → gills adnate, growing with alders and willows → *Naucoria*
    - **no** ↓ continue to top of page 23
- **yes** ↓ stalk always scaly, cap usually scaly
  - **yes** → *Pholiota*
  - **no** ↓ cap strongly hygrophanous with greasy feel
    - **yes** → *Kuehneromyces*
    - **no** ↓ cap yellow
      - **yes** → *Gymnopilus*
      - **no** ↓ cap white or dull brown
        - **yes** → *Agrocybe*
        - **no** → *Galerina*

continuation of KEY 1D

cap conical, stalk slender often tall

yes → gills free, cap sticky

- **yes** → *Bolbitius*
- **no** → cap striate
  - **yes** → *Galerina*
  - **no** → *Conocybe*

no → on wood, cap and gills golden yellow, spores bright rusty brown

- **yes** → *Gymnopilus*
- **no** → cap streaky or fibrillose, often split and with obvious umbo
  - **yes** → *Inocybe*
  - **no** → cap often slightly sticky, spores milky-coffee brown
    - **yes** → *Hebeloma*
    - **no** → spores rusty brown, cobweb-like veil on young specimens
      - **yes** → *Cortinarius*

KEY 2 Fruit body, a stalk and cap but without gills

cap with pores on the underside

no → cap with teeth

yes → *Hydnum* or *Auriscalpium*: small, cap 1-2cm growing attached to buried cones

- **yes**: *Auriscalpium*
- **no** → larger, growing in woodland litter: *Hydnum*

yes → cap very thick and fleshy, tube layer easily separable (use finger nail)

yes → surface of cap covered in coarse overlapping scales

- **yes**: *Strobilomyces*
- **no** → ring on stalk, if absent stalk granular, cap sticky
  - **yes**: *Suillus*
  - **no** → stem long and scaly, scales usually dark
    - **yes**: *Leccinum*
    - **no** → stem with a net, maybe faint (use lens)
      - **yes** → pores pink when mature
        - **yes**: *Tylopilus*
        - **no** → net white/pale, never red
          - **yes**: *Boletus or Butyriboletus*
          - **no** → pores orange/red
            - **no**: *Caloboletus (Boletus)*
            - **yes** → flesh bluing intensely, red in stem base
              - **yes**: *Suillelus (Boletus)*
              - **no**: *Rubroboletus (Boletus)*
      - **no** → continue to top of page 25

no → cap surface zoned

- **yes**: *Coltricia*
- **no** → stalk black to varying extent
  - **yes**: *Picipes (Polyporus)*
  - **no**: *Lentinus (Polyporus)*

continuation of KEY 2

↓

pores rusty red/brown,
flesh in stem yellow

no ↓ **yes** → *Chalciporus*

fruit body brown

no ↓ **yes** → *Porphyrellus*

pores orange/red

no ↓ **yes** → *Neoboletus (Boletus)*

all flesh bluing instantly

no ↓ **yes** → *Cyanoboletus (Boletus)*

cap chestnut brown,
slimy when wet

no ↓ **yes** → *Imleria (Boletus badius)*

stalk usually less than 1.5 cms diam

no → *Xerocomus (Boletus)*

yes → *Xerocomellus and Hortiboletus (Boletus)*

Hygrocybe chlorophana
Golden Waxcap

KEY 3 Fruit body, a stalk and head region NOT a cap

head region with honeycomb-like holes or with strong vertical ribs

- **no** → fruit body yellow, club-shaped
  - **yes** → fruit body slimy
    - **yes** *Leotia*
    - **no** → *Tolypocladium*
  - **no** → head green/black and sticky
    - **yes** *Phallus*
    - **no** → head region orange/red
      - **yes** *Mutinus*
      - **no** → head saddle-shaped, stalk thick and ornate or narrow and simple
        - **yes** *Helvella*
- **yes** → head region honeycomb-like or with vertical ribs
  - **no** → head cap-like, attached to the end of the stalk
    - **yes** *Verpa*
  - **yes** *Morchella*

KEY 4 Fruit body forming a bracket

brackets with pores on the underside

no → texture fleshy or rubbery
- **no**: *Stereum*
- **yes**: *Phlebia*

yes → pores bright chrome yellow
- **yes**: *Laetiporus*
- **no** → brackets blood red, fleshy
  - **yes**: *Fistulina*
  - **no** → brackets white, maybe tinged blue, soft and fleshy
    - **yes**: *Postia*
    - **no** → upper surface shaggily hairy
      - **yes**: *Inonotus*
      - **no** → brackets hard and woody

brackets hard and woody

yes → brackets very hard and hoof-like
- **yes**: *Fomes*
- **no** → irregularly shaped brackets, red/brown edge, white and crinkled
  - **yes**: *Heterobasidion*
  - **no** → brackets usually large to very large, upper surface knobbly
    - **yes**: *Ganoderma*
    - **no** → flesh corky to woody, tubes usually forming layers
      - **yes**: *Phellinus*
      - **no** → flesh softer, tubes not in layers: *Inonotus or Xanthoporia*

no → brackets with an obvious stalk
- **no** → continue to top of page 29
- **yes** → Stalk rudimentary, brackets thick and fleshy confined to birch
  - **yes**: *Fomitopsis (Piptoporus)*
  - **no** → brackets fan-shaped forming large overlapping clumps
    - **no**: *Cerioporus (Polyporus)*
    - **yes** → Pore surface bruising black
      - **yes**: *Meripilus*
      - **no**: *Grifola*

continuation of KEY 4

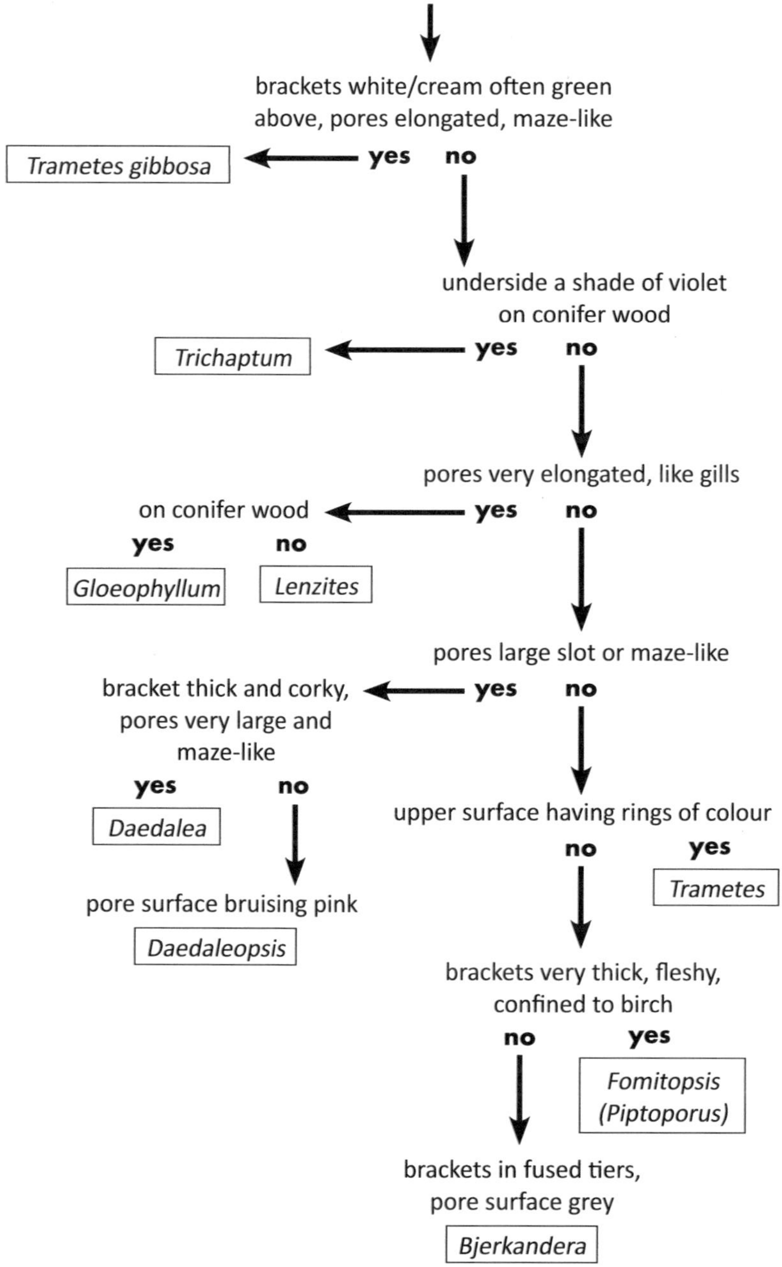

KEY 5 Fruit body spherical, hemispherical or pestle-shaped

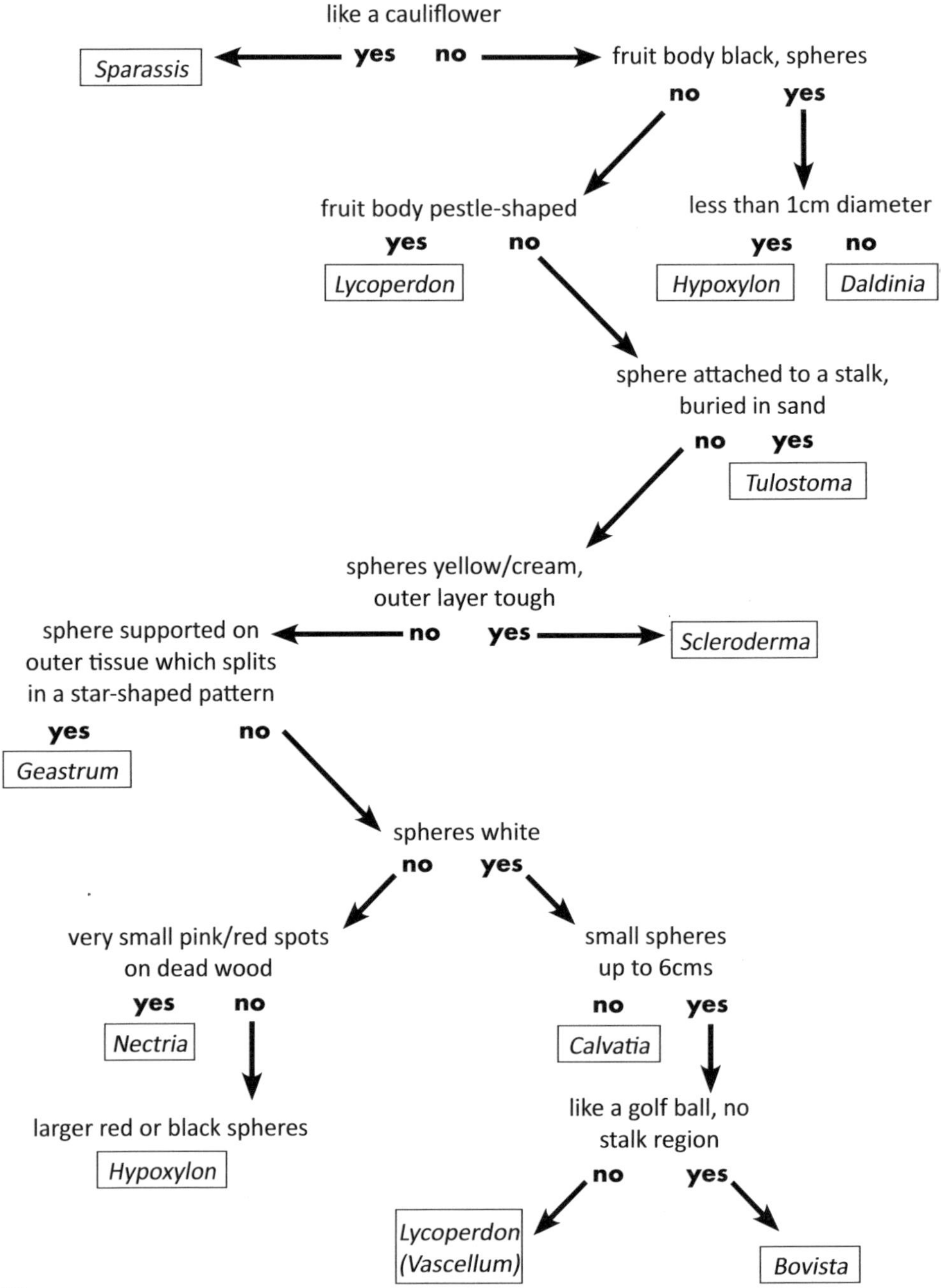

KEY 6 Fruit body forming branched or un-branched upright club-shaped structures

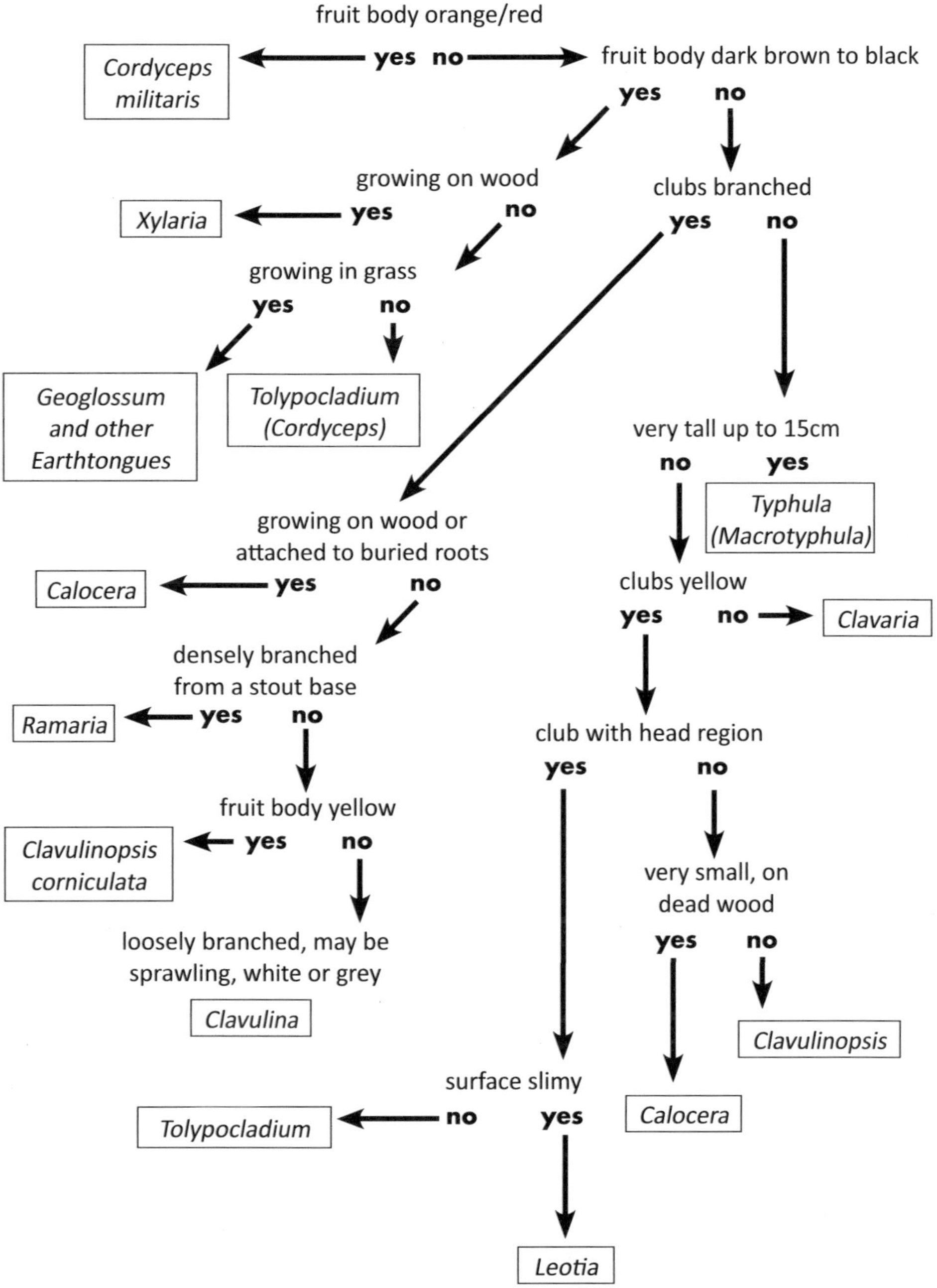

KEY 7 Fruit body cup shaped or trumpet shaped with or without a stalk

cups on a long stalk

yes → *Helvella*

no ↓

cups vivid blue-green, mycelium staining wood similar colour

yes → *Chlorociboria*

no ↓

fruit body often trumpet shaped, sometimes slit down one side

yes → fruit body dull grey/black

- **no** → *Otidea*
- **yes** → *Craterellus*

no ↓

cups orange or red

no → cups small (1-3mm), bright yellow on dead wood

- **yes** → *Bisporella*
- **no** → *Peziza*

yes ↓

growing on wood

no → fine brown hairs along the edge of the cups (use lens)

- **yes** → *Melastiza*
- **no** → *Aleuria*

yes ↓

cups about 1cm with a fringe of blackish hair around the rim

yes → *Scutellinia*

no ↓

cups larger, inner surface scarlet (in spring)

Sarcoscypha

KEY 8 Fruit body forming irregular crusts on wood

- crust black
  - **yes** → *Jackrogersella (Annulohypoxylon)*
  - **no** → surface with pores or teeth
    - **no** → surface wrinkled, commonly orange
      - **yes** → *Phlebia*
      - **no** → fruit body papery or leathery
        - **yes** → texture papery-white
          - **yes** → *Byssomerulius*
          - **no** → when cut or bruised bleeding red
            - **yes** → *Stereum*
            - **no** → crusts shades of yellow
              - **yes** → *Stereum*
              - **no** → crusts brown/purple, surface smooth
                - **yes** → *Chondrostereum*
                - **no** → *Hymenochaete*
        - **no** → pure white, resembling paint, common on elderberry
          - **yes** → *Xylodon sambuci (Hyphodontia sambuci)*
          - **no** → fruit body grey with white edge, brittle
            - **yes** → *Kretzschmaria*
            - **no** → surface cream to olive, centre fleecy, edge distinctly white
              - **yes** → *Coniophora*
              - **no** → crust firmly attached, grey, pink or lilac
                - **yes** → *Peniophora*
                - **no** → crust loosely attached, whitish cream → *Cylindrobasidium*
    - **yes** → Pores round, rusty/brown
      - **yes** → *Fuscoporia (Phellinus)*
      - **no** → pores maze-like, sometimes teeth
        - **yes** → *Schizopora*

KEY 9 Fruit body with a jelly-like or rubbery texture

- fruit body black
  - **yes** → spores black on the finger when rubbed
    - **no**: *Exidia*
    - **yes**: *Bulgaria*
  - **no** → forming small yellow orange blobs on dead wood
    - **yes**: *Dacrymyces*
    - **no** → ear shaped, often on elder
      - **yes** → *Auricularia auricula-Judae*
      - **no** → yellow or orange/yellow, folded, often brain-like
        - **yes** → *Tremella mesenterica*
        - **no** → forming rubbery brackets
          - **yes** → with small spines on the underside
            - **no** → *Auricularia mesenterica or Phlebia tremellosa*
            - **yes**: *Pseudohydnum*
          - **no** → clustered brown gelatinous lobes
            - **yes** → *Phaeotremella (Tremella)*
            - **no** → small colourless/pale yellow stalks, flattened heads on dead wood
              - **yes** → *Calocera pallidospathulata*
              - **no** → yellow orange antler-shaped on conifer wood
                - **yes** → *Calocera viscosa*
                - **no** → groups of yellow spikes on dead wood
                  - *Calocera cornea*

Russula emetica
The Sickener

Auricularia auricula-judae
The Jelly Ear

List of mushrooms and toadstools

Scientific Name	English Name
Agaricus	Mushroom
Agrocybe	Fieldcap
Aleuria	Orange peel
Amanita	Amanita
Ampulloclitocybe	Clubfoot
Armillaria	Honey fungus
Arrhenia	Moss oysterling
Asterophora	Piggyback
Auricularia	Jelly ear/tripe
Auriscalpium	Earpick
Baeospora	Cone cap
Bisporella	Lemon Disco
Bjerkandera	Smoky Bracket
Bolbitius	Fieldcap
Boletus	Bolete
Bovista	Puffball
Bulgaria	Black Bulgar
Byssomerulius	Netted Crust
Calocera	Stagshorn
Calvatia	Giant Puffball
Chlorociboria	Green Elfcup
Chlorophyllum	Parasol
Chondrostereum	Silverleaf
Chroogomphus	Spike
Clavaria	Spindle
Clavulina	Coral
Clavulinopsis	Coral
Clitocybe	Funnel
Clitopilus	Miller
Collybia	Shanklet
Coltricia	Tigers Eye
Coniophora	Wet Rot
Conocybe	Conecap
Coprinus/ Coprinellus	Inkcap
Coprinopsis	Inkcap
Cordyceps	Club
Cortinarius	Webcap
Craterellus	Horn of Plenty
Cystoderma	Powdercap
Daldinia	Cramp Ball
Daedalea	Mazegill
Daedaleopsis	Blushing Bracket
Dacrymyces	Jellyspot
Entoloma	Pinkgill
Exidia	Brain
Flammulina	Velvet Shank
Fistulina	Beefsteak
Fomes	Hoof Bracket
Fuscoporia	Cinnamon Pore Crust
Galerina	Bell
Geastrum	Earthstar
Geoglossum	Earthtongue
Gloeophyllum	Mazegill
Gomphidius	Spike
Grifola	Hen of the woods
Gymnopilus	Rustgill
Gymnopus	Toughshank
Hebeloma	Poisonpie
Helvella	Saddle
Heterobasidion	Root Fomes
Hydnum	Hedgehog
Hygrophoropsis	False Chanterelle
Hygrocybe	Waxcap
Hygrophorus	Woodwax
Hymenochaete	Crust
Hyphodontia	Whitewash
Hypholoma	Brownie
Hypoxylon	Woodwart
Inocybe	Fibrecap
Kretzschmaria	Brittle Cinder
Kuehneromyces	Woodtuft
Laccaria	Deceiver
Lacrymaria	Weeping Widow
Lactarius	Milkcap
Laetiporus	Chicken of the Woods
Leratiomyces	Reolead Roundhead
Leccinum	Bolete
Lentinellus	Cockleshell
Lenzites	Birch Mazegill
Leotia	Jellybaby
Lepiota	Dapperling
Lepista	Blewit

(continued inside back cover)

List of mushrooms and toadstools

(continued from inside front cover)

Scientific Name	English Name
Lichenomphalia	Navel
Lycoperdon	Puffball
Lyophyllum	Domecap
Macrolepiota	Parasol
Macrotyphula	Club
Marasmius	Parachute
Megacollybia	Whitelaced Shank
Melanoleuca	Cavalier
Melastiza	Orange cup
Meripilus	Giant polypore
Morchella	Morel
Mutinus	Dog Stinkhorn
Mycena	Bonnet
Naucoria	Aldercap
Nectria	Coral spot
Neobulgaria	Jellydisc
Otidea	Ear
Oudemansiella	Porcelain fungus
Panaeolus/ Panaeolina	Mottlegill
Panellus	Oysterling
Paxillus	Roll Rim
Peziza	Cup
Phallus	Stinkhorn
Phlebia	Wrinkled crust/ Jelly rot
Phellinus	Cork Bracket
Pholiota	Scalycap
Pleurotus	Oyster
Pluteus	Shield
Polyporus	Polypore
Psathyrella	Brittlestem
Pseudohydnum	Jelly tooth
Psilocybe	Brownie
Ramaria	Coral
Rhodocollybia	Toughshank
Rhodotus	Wrinkled Peach
Rickenella	Mosscap
Russula	Brittlegill
Sarcoscypha	Scarlet Elf cup
Schizopora	Split Pore Crust
Scleroderma	Earthball
Scutellinia	Eyelash
Sparassis	Cauliflower
Stereum	Crust
Strobilurus	Conecap
Strobilomyces	Old man of the woods
Stropharia	Roundhead
Suillus	Bolete
Tapinella	Rollrim
Tremella	Brain
Trichaptum	Bluepore
Trichoglossum	Hairy Earthtongue
Tricholoma	Knight
Tricholomopsis	Plums & Custard
Tubaria	Twiglet
Tulostoma	Stalk Ball
Tylopilus	Bitter Bolete
Vascellum	Puffball
Verpa	Morel
Volvariella	Rosegill
Xanthoporia	Alder Bracket
Xerula	Rooting shank
Xylaria	Candlesnuff / Fingers

Fungus Conservation Trust

The Trust as a registered charity exists to help people like you develop your interest in the fascinating world of mushrooms and toadstools. It provides opportunities to foray with local experts in your area and is particularly welcoming of those developing an interest in fungi. Members are all amateur enthusiasts who want to broaden their interest and help with conservation.

Contact The Fungus Conservation Trust:
Harveys, Alston, Axminster, Devon, EX13 7LG, UK; enquiries@abfg.org; 01460 221788.

Recommended books to use with this key

GENERAL

Bonn, M., *The Mushrooms and Toadstools of Britain and N. W. Europe* (1987).

Buczacki, S., *Collins Fungi Guide* (2012).

Courtecuisse, R., and B. Duhem, *Mushrooms and Toadstools of Britain and Europe–Collins Field Guide* (1995).

Jordan, M., *Encyclopedia of Fungi of Britain and Europe* (2004).

Kibby, G., *Mushrooms and Toadstools of Britain & Europe, Vol. 1* (2017), vol 2 (2020).

Laessoe, T., *The Mushroom Book* (1996).

Laessoe,T., and Petersen, J.H., *Fungi of Temperate Europe* (2019)

Mcadam, A., *First Steps in Fungi Identification* (2010).

Pegler, D. N., *Field Guide to Mushrooms and Toadstools of Britain and Europe* (1998).

Phillips, R., *Mushrooms* (2006).

Sterry, P. and B. Hughes, *Collins Complete guide to British Mushrooms & Toadstools* (2009).

MORE SPECIALISED

Breitenbach, J. and F. Kranzlin (eds), *Fungi of Switzerland* (1984).

Kibby, G., *The Genus Russula in Great Britain* (2007).

Kibby, G., *The Genus Agaricus in Great Britain* (2011).

Kibby, G., *British Boletes* (2011).

Knudson, H. and J. Vesterholt (eds), *Funga Nordica* (2012).

Legon, N.W. and A. Henrici, *Checklist of the British & Irish Basidiomycota* (2005).

Moser, M., *Key to Agarics and Boleti* (1983).

Outen, A. and P. Cullington, *Key to Species of Inocybe* (2012).